CALM

JOHNS HOPKINS
UNIVERSITY PRESS

AARHUS UNIVERSITY PRESS

calm

IBEN HAVE

CALM
© Iben Have
and Johns Hopkins University Press 2023
Layout and cover: Camilla Jørgensen, Trefold
Cover photograph: Anders Bach
Publishing editor: Karina Bell Ottosen
Translated from the Danish by Heidi Flegal
Printed by Narayana Press, Denmark
Printed in Denmark 2023

ISBN 978-1-4214-4606-6 (pbk)
ISBN 978-1-4214-4607-3 (ebook)

Library of Congress Control Number: 2022938081

Special discounts are available for bulk purchases of this book. For more information, please contact Special Sales at specialsales@jh.edu.

Published in the United States by:

Johns Hopkins University Press
2715 North Charles Street
Baltimore, MD 21218-4363
www.press.jhu.edu

Published with the generous support of the
Aarhus University Research Foundation

Purchase in Denmark: ISBN 978-87-7219-192-8

Aarhus University Press
Finlandsgade 29
8200 Aarhus N
Denmark
www.aarhusuniversitypress.dk

CONTENTS

IMPRESSIONS AND EXPRESSIONS

CLOISTERED LIVING

In the interest of disturbing the reader as little as possible, this volume really ought to consist of nothing but blank pages. On second thought, a reader might well become uneasy at encountering an absence of words to read in a book such as this.

While it might provide a moment's pause, I suspect the void would rapidly fill with unease, mingled with irritation and disappointment. No: In order to foster calm it is surely best to lead you, as a reader, along the path of the familiar format of the book, enabling you word by word, minute by minute, to let your mind follow my meanderings on a single concept: calm.

These reflections are taking shape at San Cataldo, a centuries-old convent on the Amalfi Coast in Italy, where I have come to find peace and quiet. In this tranquil setting I can write, really delve into my topic and think at length. Some would say it's too calm here, others that it's not calm enough, but for me this retreat is the perfect place to work.

My study cell is on the lower floor, next to the bathrooms, where handsome shoulders of bare bedrock jut through the white walls. Perched 500 metres above sea level, the restored convent overlooks the bay and the town of Ravello, across the gorge. My mind soars as I gaze at the mass of weathered mountains whose solemn, unmoving aspects have defined this landscape since time immemorial.

The high-ceilinged rooms are decorated in gentle shades of grey and green – except for the bedspread, with its jarring contrast of dark blue and snow-white squares. I've stashed it in the closet and replaced it with a green-patterned scarf of my own.

It is April, so few flowers are blooming in the cloistered garden. The exceptions include a profusion of lavender-blue wisteria, a bed of white lilies of the valley and the occasional deep blue iris. Little noise penetrates the massive walls, which also prevent the spring warmth outside from reaching me. For days my cell stays chilly with winter's cold. I have no choice but to turn on the electric fan heater, but my ears soon get used to its constant hum. I finally stop noticing it at all, and remain calm and unperturbed.

I am here with 11 other guests. Each of us is working individually on our own project. This makes mealtimes an important and carefully choreographed part of our social lives. The first thud on the gong means 'ten minutes to go', either till lunch at 1 o'clock or dinner at 7.30. The second thud means 'take your seats, please'. This way we don't have to keep track of the time ourselves.

The matron changes the seating each evening to vary

the conversation. The kitchen staff bring large platters to either end of the oval dining table. We pass the platters, always to the right, and they are collected at the opposite end of the table. At first these routines seem artificial, but over time they create rhythm, predictability and a sense of calm.

San Cataldo offers a blend of visual, auditory and social calm that provides me with the optimum conditions for working creatively and immersively. I am convinced that my thinking is better, deeper and more focused when I am not interrupted by new, urgent tasks, and pings and notifications from my phone, email and calendar app, all of which tends to make me feel like I never quite finish what I set out to do.

I normally work from my office in Denmark, as a researcher at Aarhus University, but by abandoning my routines and travelling to San Cataldo, I have staged the onset of calm, and I anticipate it. We all know how often the world is shaped by the way we consciously think about it. Sitting here, 1,750 kilometres from my home, I imagine how a thousand years' worth of residents, all cultivating calm and contemplation in this place, have saturated the very fabric of its walls.

San Cataldo became a Danish-owned work retreat in 1924, but for 900 years before that it was a working convent where Catholic nuns immersed themselves in prayer as a community, following the meticulously detailed 'Rule of Saint Benedict', which was set out in the sixth century by Benedict of Nursia at a monastery he founded between Rome and Naples. Typically, nuns were unmarried daughters of wealthy local families, many from the nearby town of Scala. A common saying in Italy was *Aut virum, aut murum*, 'either a man or a wall'; if a girl did not find a man, her only alternative was a convent.

In fact, nuns were married, not physically to any mortal man but metaphorically to Christ. The bishop would symbolically place a 'ring of faith' on a novice's finger to witness her commitment as the 'wife of God'. With this she would renounce her former life, have her hair cut short and replace her regular clothes with a simple nun's outfit. This black and white 'habit' was the same for all nuns, with little to arouse the senses. All were equal in the eyes of the Lord.

The life of the convent was stable and minutely regulated, and it remained largely unchanged for centuries. This way the nuns did not constantly have to make decisions, adapt to new situations and people, or be 'ready for change'. Like modern-day guests at the retreat, the nuns lived a quiet, cyclical communal life with little contact with the outside world.

During the Reformation, in the 1500s, Protestants in Northern Europe spread rumours of impiety and debauchery in the Catholic monasteries and convents of Southern Europe. These rumours soon reached the Catholic Church, thanks to a new technology – printing, invented in the mid-1400s – which enabled rapid communication over much greater distances than before.

The Catholic Church reacted in 1545 by introducing 'enclosure', a type of enforced isolation that was also applied at San Cataldo, preventing the nuns from leaving their convent or receiving any visitors. Under the Rule of Saint Benedict they had already given up owning any personal belongings, and the Rule's directives also included certain times for silence and speech. Breaches were punished, and offenders might be ordered to lie outstretched on a cold floor for hours.

This strict lifestyle was meant to enable the nuns, through contemplation, immersion and prayer, that is, to turn their full attention to God. 'Contemplation' is still a key word at San Cataldo, but now the old convent enables musicians, artists and researchers like me to benefit, briefly, from these peaceful surroundings and turn our full attention to our work.

BALUBA AND TAKETE

Take a good look at the word 'calm' and think about how it is pronounced. Let it fill your mind. Doesn't it almost seem to instil what it denotes? Now say it out loud. The initial 'k' is admittedly an abrupt start, but it only involves the back of the tongue and the soft palate, leaving the open lips still. It moves into a neutral 'ah', like a tiny sigh, with the silent 'l' barely hinted before the lips close into an 'm' that turns the oral cavity into a resonating space. Now write the word in small letters. Especially in cursive handwriting, it has a smooth feel to it. No 'i's to dot; no 't's to cross.

The English word 'calm' is almost as soothing as its Danish counterpart – *ro* – which rolls easily off my native tongue. Short and sweet, round and rounded, it begins at the very back with a guttural 'r', then moves to a full-bodied 'o' and ends with a soft glottal stop, as many Danish words do. It is worth noting, however, that 'calm' and *ro* are not fully synonymous. One big difference is that 'calm' is a noun, an adjective and a verb, whereas *ro* is only a noun in this sense, but it looks and sounds exactly like the unrelated and very active Danish verb *at ro*, which means 'to row a boat'.

Looking a bit further afield, a Dane or German might use their 'calm' word as a command to quiet a large assembly – *"Kan vi så få ro?!"* or *"Ruhe, bitte!"* – whereas the English and French words seem better suited to calm or soothe on an interpersonal level – "Calm down now, sir" or *"Calmez-vous, monsieur"*.

In Eastern philosophy, the most sacred mantra in Hinduism and Buddhism – *Om* or *Aum* – is a primordial sound that unites the physical and spiritual aspects of life. The Sanskrit symbol for *Aum* – ॐ – and its sound both have sensory qualities reminiscent of 'calm' and *ro*.

These may seem like mere musings, but when it comes to words, soft visual shapes and neutrally open or rounded sounds can suggest an abstract experiential quality that is perceived across different modes of sensory expression and perception. Back in 1929, the German-American psychologist Wolfgang Köhler conducted an experiment later reproduced, with various modifications, by others. They all confirmed a link between certain visual forms and the pronunciation of special words constructed for the experiments.

Köhler showed his subjects two figures: one a curvy shape, like a child's outline of a flower in bloom, and the other an asymmetrical, pointy star shape. Then he asked them which figure they would call *takete* and which they would call *baluba*. A full 98% assigned *baluba* to the curvy shape and *takete* to the pointy shape.

Köhler's findings were later dubbed 'the bouba / kiki effect', based on the words used in 2001 for a similar experiment. The researchers found that people respond uniformly across different language cultures and age groups, demonstrating a universality in the way humans link visual and auditory forms. In other words, people's perceptual experiences are not based on culture alone, but are evidently rooted in a deeper pan-human phenomenon.

Science has offered various explanations of this phenomenon. Some researchers say that we understand sensory impressions based on the way our physical bodies exist and act in the world. Working with this explanation, scientists have studied how our physical, bodily experiences create certain basic structures and patterns in the brain, known as 'mental schemata', which control our perception and understanding of sensory impressions.

Human beings collect bodily experiences with stillness and motion, smooth and prickly surfaces, soft and sharp objects, cold and heat as we move around in the world touching knives, balls, hedgehogs and kittens. These sensory recordings shape our mental schemata, which in turn shape how we perceive words like *takete* and *baluba*. Today, several studies of the human sensory apparatus also point out that distinguishing among sensory inputs of a single, isolated type is not meaningful. Instead, they recommend that we understand types of input more holistically, as elements affecting one another as we experience them.

The Canadian writer Margaret Atwood uses intensely evocative sensory amalgamations to render the mental state of her characters. One example is found in her futuristic dystopian novel *The Handmaid's Tale* from 1985, which gave rise to a TV series launched in 2017: "You can wet the rim of a glass and run your finger around the rim and it will make a sound. This is what I feel like: this sound of glass. I feel like the word *shatter*." This is how the servant-slave Offred describes her own state, sitting in her room after taking part in a bizarre ceremony to potentially breed a child for the Waterfords, the married couple whom she serves. Forced to suppress her own sexuality, she feels attracted to Nick, the household's chauffeur, even while longing for her vanished husband.

A duplication of Köhler's experiment using the spoken words 'calm' and 'frantic' would probably confirm the bouba/kiki effect. Should any readers wish to try this at home, they will need subjects who can disregard the meaning of the words and focus only on the sound. And, obviously, the same connotations do not apply to all near-synonyms of 'calm/*calme*', or '*ro/Ruhe*' – as evident in the words 'tranquillity' and 'quiet'.

For some people, experiencing cross-sensory perception is an automatic reaction, technically referred to in psychology as 'synaesthesia'. Some psychologists regard synaesthesia as a medical condition. Predisposition differs greatly: In some the reaction occurs readily, in others not at all. However, most people are able to experience mild synaesthesia, much like an instinctive bouba/kiki reaction.

I sometimes perceive high-pitched, discordant voices as yellow, or the flavour of a good red wine as the timbre of a cello. When I describe a cursive, written 'calm' as smooth, or a spoken *ro* as rounded, I am drawing on synaesthetic perceptions and bodily experiences I probably share with many others. When you meet new people, there is no guarantee that Bea or Bob will be less vivacious than Vicky or Ricky, but even before we get to know them, the letter combinations in their names signal certain sensory traits.

ABSENCE OF MOTION, ABSENCE OF SOUND

The Danish word *ro* is deeply rooted in all the Nordic languages, also appearing in the Icelandic sagas, which tell stories of the country's Viking inhabitants who lived about a thousand years ago. The noun *ro* has three main meanings: a state of physical rest, without motion; a relaxed mental state, free of unease; and an absence of sound. *Ro* is also an inherent part of *hygge*, a Danish concept with no precise equivalent in English: a sort of simple, pleasant, unhurried state of being with others, or alone, which rests in itself.

The English word 'calm', in its various uses, embodies the same three aspects as the Danish *ro*: The adjective means not excited, nervous or upset, or not in motion. The noun and verb relate to being or becoming quiet, peaceful and relaxed, often as opposed to a preceding time or situation that was not calm.

Many English idioms use words other than 'calm', whereas Danish idioms often build on the versatile noun *ro*. But both languages define the phenomenon by its antithesis – by what it is *not*. An absence of movement, tension and sound creates calm. This is correct, in a scientific sense, if you regard calm as a detectable, quantifiable phenomenon. But if we weave human experience into the equation, the very presence of movement and sound can create calm – as expressed in the definitions, Danish and English, that reflect the absence of mental unease.

We can find calm in a sound, like raindrops on a window pane, or New Age music, or the deep voice of the beloved Danish children's television host Thomas Winding reading a story about Pippi Longstocking. If used deliberately, calming sounds can produce the sort of balanced mental state like many yoga instructors, mine included, seek to create by beginning sessions with a collective *Aum*, to make us present in the here and now. Inner calm can also be found in physical movements such as running or touching, or in looking at a decorative mobile – which Danes call an *uro*, 'a restlessness'.

The study of calm does not belong in any single academic field. As a broad daily concept with no firm definition, 'calm' takes on meaning depending on how we use it. It is a situational state that we can ask others to help create around us – so we can sleep, eat or work – and it is also a mental state we can personally strive to attain. Related words are tranquillity, peace and quiet, balance, harmony, contemplation and concentration, to name a few. My research field is sound and media, which may seem a bit absurd, given that these aspects of our lives are often blamed for being exceptionally noisy and disturbing. What intrigues me, however, is exploring how we can find calm using sound-based media, digital technologies and our own bodies and minds.

ROUTINES AS A 'HELPDESK'

You may have chuckled your way through a YouTube video called "Medieval Helpdesk", uploaded in 2007 by the Norwegian Broadcasting Corporation, NRK, and performed in lilting Norwegian, but accessible with English subtitles. It opens with Brother Ansgar, a well-known figure from Dano-Norwegian history, sitting at his desk in a candle-lit monastery cell, despondent after failed attempts to use the cutting-edge technology of his day: the spine-bound manuscript. He finally accesses the 'helpdesk' – in the form of another monk, who enters his cell.

Brother Ansgar, whose tech savvy stops at scrolls, vents his frustrations with the new-fangled device in front of him. The helpdesk monk then demonstrates, with exemplary patience, how Ansgar can open and close the tome, leafing from right to left and reassuring him that the text runs on, from page to page, and will remain stored in the book, even after they close it. Comical, and relatable.

Like Brother Ansgar, we always feel alienated when

dealing with new technology. Fortunately, we can practice and remember routines. I point this out because finding calm has a lot to do with habits, routines and predictability. Imagine having to figure out how to use a book every time you picked one up. Frustration guaranteed! After 1,500 years of using bound books we are quite comfortable with them. What frustrates us today is information and communication technology, with its endless stream of new apps and operating systems.

Embedded bodily knowledge and familiar routines are reassuring at a micro and macro level. I know what I'm having for breakfast, which route I will cycle to work, where to find my office and how to unlock its door. We are equally comfortable with familiar music, stories and holiday traditions. Now imagine you really had opened this book to find only blank pages. You would probably feel insecure; you wouldn't know what to do with it. The hard truth is that in the real world, we don't have a helpdesk to handle every unknown eventuality.

HOME, SWEET HOME?

"Silent night, holy night. All is calm, all is bright." These words open a Christmas song that originated in Austria in 1818 and soon appeared in translated versions across Europe, and later around the world. In Denmark it is also a holiday favourite, striking a solemn but joyful chord in the Christian tradition. UNESCO has even declared "Silent Night" a part of humanity's intangible cultural heritage.

The composition itself expresses peace and quiet in a

slowly rocking 6/8 time signature, with simple, smooth phrases that are repeated or sequenced. The energy rises surprisingly in the culminating phrase "Sleep in heavenly peace!", but comes to rest just as quickly, with the final phrase descending to the melody's last and deepest note.

I have sung this song in Danish countless times, and in English, German and Italian as well. I have sung it to myself, with friends and family around the Christmas tree, and in concert with many different choral groups over the years. As a choir director I have even led others in singing it, so I can reasonably claim to be intimately familiar with its pitches and rhythms.

Music research has documented that just one wrong note in a melody we know well creates a disturbance, arousing our nervous system and making us more alert. Eurovision Song Contest melodies and other kinds of pop music notoriously exploit this phenomenon, raising the key by half a tone near the end for greater intensity – to perk up the audience, just in case their minds had begun to wander.

Unexpected, foreign or irregular sounds and harmonies make us uneasy, but steady rhythms and constant, predictable sounds and soundscapes give us a sense of control and therefore of calm. That's why the fan heater in my cell at San Cataldo doesn't bother me. Habit also decides which types of music we consider noisy. My husband, for instance, finds heavy metal music soothing. When I ask him why, he replies that it is "recognisable".

And yes, he is intimately familiar with the genre, the mode of expression and the vinyl he puts on his record player.

The same goes for stories, most of which follow certain well-known narrative patterns, notably the 'home-away-home' pattern. The Danish storyteller Hans Christian Andersen's famous tale of *The Ugly Duckling* from 1843, which is ostensibly autobiographical, begins with pastoral bliss. Mother Duck is bursting with pride over her special duckling. But after they move into the clamorous poultry yard, the ugly duckling, constantly pecked and bullied, is forced to flee. He tries his best to get by in the world, but all those he meets finds him ungainly – the eternal misfit. Only when he finally, humbly, meets the great swans does the ugly duckling make peace with himself, finding happiness, kinship and his true home.

Just like *The Ugly Duckling*, narratives that follow this pattern usually have a happy ending, leaving us relieved and reassured. Stories that surprise us with an unresolved ending leave us feeling uneasy, and we step away from the film or book yearning for closure. The 1968 cult horror film *Rosemary's Baby* is an excellent example, which I may well spoil for you by revealing that Rosemary, played by the American actress Mia Farrow, gives birth to the son of Satan himself and, driven by maternal instinct, ultimately decides to care for her offspring – gently rocking the newborn creature's cradle.

But 'home sweet home', harmony and happy endings yield little progress or innovation. As the French economist Jacques Attali has pointed out, social change always involves disruption and noise. Art and evolution arise when expectations are challenged. It is surprises, missteps and obstacles that bring us new knowledge and new perspectives on the world.

THE RESTLESS HUMAN MIND

FREE BUT UNEASY

People have always sought islands of calm. We need peace and quiet, especially to fall asleep. Like clean air and water, calm is a basic human need. We depend on it, and we are responsible for preserving and protecting it. But calm has become a rare commodity. Modern societies are full of noise – literally and figuratively speaking – and noise triggers the release of the stress hormone cortisol to a part of our brain called the prefrontal cortex. An excess of cortisol for prolonged periods of time can be so stressful that it makes us ill.

The World Health Organization, WHO, classified stress-related illnesses as one of the heaviest health burdens of 2020. That is why, nowadays, calm – peace, quiet, stillness – is a coveted resource that modern, busy people are increasingly cultivating, seeking out and incorporating into their identity.

In traditional societies, which build on heritage and family ties, a carpenter's son almost inevitably becomes a carpenter, and a fisherman's son a fisherman. Girls and women do not need to worry about their roles either, being predestined to work in the home or the family business.

Around the mid-1800s, many traditional societies developed into modern ones, and people were gradually liberated from the constraints of convention. When our path in life is no longer set out by a God, destiny or tradition, and when life and death are no longer attributed to a God's will, we must each find our own way. This results in unease and insecurity, and while personal freedom has brought many benefits and opportunities, the costs of individualisation include uncertainty and a lost sense of belonging.

We begin to cultivate ourselves, preoccupied with our looks, our food, our friends on Facebook, and the number of 'Likes' our new profile picture gets, comparing ourselves to our friends. Yet perpetual identity construction and optimisation foster insecurity, uncertainty – 'un-calm', like the motion of an ever-shifting, restless mobile. On top of that, our digitised societies have accelerated the pace of most things in life.

Individualisation has turned our search for islands of calm, harmony and balance into an individual project. The nuns of San Cataldo could lay their lives in the hands of an almighty God, finding existential calm in their faith and trust in the Lord. As for the rest of us, God only knows what we do without Him.

SAVORING SLOWNESS

The guests at San Cataldo eat simple meals made from local

produce, much of it from the convent's own garden. This April, during my stay here, the stores of food seem to be running low on fresh produce, except for beans and Swiss chard, which we enjoy in a variety of dishes. The wine, made from the convent's own grapes, is kept in old barrels outside the kitchen, alongside large, eye-catching flagons of homemade vinegar. And to be fair, the pantry is still amply stocked with bottled tomatoes, boxed cheeses and generous bunches of dried herbs.

The food's journey from garden to table is slow, and guests are asked to lend a hand when grapes or tomatoes need picking. The cook is in no hurry, and neither are the guests. The food is unpretentious but delicious, and the modest helpings are just enough to satisfy our hunger. For the nuns, these practices were a matter of survival and perhaps of giving thanks for the Lord's blessings. Today, they would qualify as 'slow food' – part of a new global trend.

The Slow Movement, a concept launched by the journalist Carl Honoré in 2004, has inspired 'slow tourism', 'slow fashion' and even 'slow media'. A semi-meditative genre called 'slow TV' became a viewer sensation with an NRK broadcast of a seven-hour train journey on the Bergen Line in 2009. The Norwegian programme is one continuous shoot of the train chugging its way from Bergen to Oslo, past fjords, mountains and woodlands, as seen from the train driver's seat. The British radio station BBC Radio 3 has also enjoyed considerable success with a series called *Slow Radio*, branded with a snail-inspired logo as an "antidote to today's frenzied world" under the motto "it's time to go slow". One element all these slow movements share is their focus on what and how their followers consume, rather than how much or how quickly.

FROM PRANAYAMA TO AMYGDALA

The Danish television drama series *Herrens Veje* – known in English as *Ride Upon the Storm* – originally aired in 2017 and introduced the character Johannes Krogh, a controversial pastor in the Danish Lutheran church, and his family. In the first season the restless elder son, Christian, travels to Nepal, where he meets a Buddhist monk. The monk, whose Nepalese name means 'compassion', helps Christian let go of his repressed anger towards his father, Johannes, and focus on finding calm in his own life by "learning how to breathe", as the characters in the series put it.

Back in Denmark, Christian starts a company called Open Mind, giving talks and coaching others on Buddhist teachings adapted for Western consumers. His message, although well intentioned, becomes obscured by superficial commercial management logic, which drowns out his original goal of helping others find calm. Critics have a name for such Westernised neoliberal versions of Asian religious practices: 'McMindfulness'.

Today, many do as Christian did in the series, finding and cultivating calm through meditation or its modern Westernised version, 'mindfulness'. Both keep the body still while the mind works to maintain focus on breathing or on registering and controlling thought activity. Meditation allows us to practice keeping negative feelings like worries and envy at bay as they clamour to get attention.

David Hume, an eighteenth-century Scottish philosopher, believed that humankind's morality lies in our ability to work with our own emotional lives. Hume distinguished between "calm and violent passions" and referred to one of the calm passions as "fellow-feeling", essentially empathy, which we can use to control the violent passions. But keeping the former in the driver's seat takes a great deal of self-control, given that the latter are constantly trying to drown them out.

Science now acknowledges that daily mindfulness training reduces brain activity in the areas associated with repetitive thinking, worry and self-criticism. Laboratory experiments have also shown that calm breathing positively affects the parasympathetic nervous system, also called the 'rest and digest system'. I imagine the Buddhist monk from Nepal would have used a Sanskrit concept and told his new friend Christian to practice *pranayama* – had they not both been characters in a prime-time drama series.

Studies have also shown that mindfulness and meditation physically increase the thickness of the cerebral cortex in the seahorse-shaped brain structure known as the *hippocampus*, positively affecting learning and memory. Meditation also reduces activity in the brain's emergency-response centre, the *amygdala*, which processes fear, anxiety and stress.

THOUGHT CONTROL

As I sit in my cell at San Cataldo writing these lines, I am interrupted by a recurring thought that springs to mind: I have a piece of chocolate in my drawer that I would really like to eat. As I dedicate increasing amounts of energy to not giving in to this thought, my urge to devour the chocolate grows. It must have been easier for the cloistered nuns, for whom even the smallest temptations remained outside the convent's walls.

In this distracted state I strive to discipline my thoughts with reason and not let my urge disturb the work I have promised myself I will complete. Besides, common sense tells me it is not wise to eat chocolate just half an hour before lunch.

In his 2017 book *Why Buddhism is True*, the American science journalist Robert Wright theorises that meditation can equip us to live in societies full of constant temptations and interruptions. Evolution has coded humans to fulfil our basic needs as quickly as possible, but in societies where we no longer have to hunt, gather or even prepare our own food, highly processed industrial foods tempt us and are easily accessible. This means our encoded mechanism can easily go off kilter or out of control.

If I crave chocolate, I can easily get some, but the satisfaction is brief. When the chocolate is gone, I want more. These constant urges distract us and obstruct concentration. Today, because we have more time, more money and a media landscape that constantly offers quick fixes and instant gratification, we train our minds to give in to our urges and cravings for satisfaction here and now, whether what we 'need' is chocolate, a new dress, to binge-watch our favourite TV series or to check the latest posts on Instagram. In the short term this calms us down, but in the longer term it can lead to restlessness and mental disquiet.

Wright describes meditation as a cultural method that enables us to adapt to a noisy world. Until now, the evolutionary development of our species has not been controlled by individuals needing to be happy and well-balanced to survive. However, humans may have reached an evolutionary phase where an upsurge of depression, anxiety and stress calls us to actively seek calm and meditative thought control to avoid ruin. Meditation can teach us how our emotions constantly affect our thoughts; negative emotions like anger and fear are the greatest challenge for anyone trying to achieve and maintain peace of mind.

The power of human emotion can be used strategically to influence people's opinions and decisions. The British IT company Cambridge Analytica exploited this when it infamously collaborated in the UK with the Leave.EU movement, and in the US with Donald Trump's 2016 presidential campaign. Using data from Facebook and other sources, they were able to target and send more or less truthful campaign propaganda to millions of people, tailored to affect individual voters' emotions based on their personal lifestyle and preferences.

This was possible because, as humans, we are

unconsciously and often uncritically attracted to opinions we already have, and blind to those we do not share. This mechanism is not always beneficial in democratic systems built to be fuelled by objective facts and arguments. If meditative practices can create inner calm, perhaps they can also increase our awareness of how our emotions naturally affect our thoughts, opinions and actions, which could ultimately make us more critically aware as citizens – perhaps enough to make Internet trolling bounce, like skipping stones, across the smooth surface of our unruffled minds.

AUMMM ...

Twice a week I attend a yoga class at the Aarhus University gym, where I see some of my colleagues from a whole new angle. During the physical exercises we focus on ourselves, trying to blend our movements, breathing and thoughts into a harmonious whole. Twice during each class, we turn our attention outwards, towards each other, both times using sound. The first time is when we begin the class by intoning a collective *Aum*, seeking to tune into each other and the resonance in the room. The second time is near the end of the class, during relaxation. As we lie with our eyes closed, the yoga instructor strikes full-bodied tones with felt-covered wooden mallets on variously sized bronze singing bowls. The materials and shapes of bowls and mallets meet, creating a soft, reverberating ringing that slowly subsides.

I vividly remember the first time I lay there hearing – or

rather, sensing – the deep vibrations from the bowls. I do not subscribe to the spiritual beliefs about what singing bowls do, but I can feel the physical resonance their sounds create in the floor, in the room, in my body.

'Resonance', as a word, is related to 'resound', meaning 'to sound again', and physically it is a phenomenon that arises when two systems or objects begin to oscillate together. Acoustic resonance happens when someone strikes a dinner gong, for instance, or when sound vibrations from strings strike the hollow body of a violin or a guitar and the waves spread to the material and the surrounding space. That is why a particular instrument's shape and substances are decisive for the quality of its sound. The same goes for the human voice, which is distinctive because of a person's body and vocal resonance cavities.

Resonanz is also the title of a book by the German sociologist Hartmut Rosa, which appeared in German in 2016. He uses the concept of 'resonance' as a metaphor for a way of being in contact with the world, which is in contrast to daily lives and societies typified by distance and competition. Rosa believes these trends are accelerating in modern societies, which are now in overdrive. Our globalised, digitised societies are caught up in a sort of expanded freedom, where we are constantly obliged to reinvent ourselves to avoid losing momentum and collapsing. We are forced to run faster and faster just to stay in place, as Rosa puts it.

In Rosa's view, however, the solution is not necessarily

to slow the pace and cut the power, as Slow Movement advocates would do, since acceleration is only a bad thing when it leads to alienation. Instead, he says, we need to re-create our connection or resonance with the world, the absence of which is making us feel alienated. The sense of hearing is absolutely crucial here, for as a human being one can only achieve resonance by 'listening' to people, objects or the context in which one is embedded.

He describes the conditions for the occurrence of resonance as social and political, and consequently the aim must be to create a social and societal framework that enables people to experience resonance. A capitalist society, Rosa says, which focuses on optimising and competing and burdens its citizens with constant time pressure, will counteract a resonance-based relationship between us and the world around us, thereby also counteracting a good, meaningful life. This societal perspective and responsibility are often forgotten in mindfulness and cognitive therapy, which are, conversely, based on us working, often in isolation, to 'get our act together' and get well again.

Rosa distinguishes between three types of resonance, each one an 'axis' along which we can become happier. The first axis consists of horizontal resonance, which arises in interpersonal relations such as love or experiencing nearness or rapport with others. The second axis is diagonal, where resonance can arise between people and material objects, as in the way the carpenter relates to the wood, or the football player relates to the ball. The third axis is vertical, and it concerns the relation between ourselves and the deeper layers of our existence, which can arise in our encounters with religion, history, art, philosophy or nature.

In many ways, Rosa's concept of 'resonance' resonates well with the understanding of calm I am trying to convey in this book. Instead of delimiting and defining 'calm' as a concept, I am telling the reader about calm as a phenomenon or a state we aspire to achieve, through sounds and actions, among other things. Resonance and calm can arise through the ways in which we, as human beings, relate to ourselves and our surroundings.

When we experience Rosa's three types of resonance, we get a sense of being connected with the world around us, and this brings us calm. Not in the sense that we are immobile, immutable and rational, but in the sense that we are vibrating, pulsating, shifting and changeable. Rosa says that when we resonate – with a loved one, a podcast, Denmark's wild West Coast where Jutland meets the North Sea, or even just a good idea – it changes us as people. This makes the concepts of 'calm' and 'resonance' mutually conditional, with their interaction calling upon human intonation: A person who does not experience resonance in Rosa's sense of the word will remain in a state of introverted unease – and a society that leaves no space for calm will prevent resonance from arising.

SOUNDING OUT CALM

CALMING NOISE

Digital media are noisy, vying for our attention with constant notifications and addictive algorithms. Paradoxically, the digital world can give us access to peace too, thanks to apps with names like Calm, Chill and Relax, Calming Gong and even White Noise that help us create our own little islands of calm, whenever and wherever we want.

Users of the Calm app first see a placid lake bordered by evergreens against a background of snowclad mountains, set to sounds of trickling water and twittering birds. The menu at the bottom of the screen offers various options, including bedtime stories with titles like "Waterfall", "Gratitude", "Calm Airways" and "Cricket Explained", read slowly by soothing voices that sometimes also instruct the listener on how to relax.

Noteworthy titles include "Once upon a GDPR", which features the legendary BBC continuity announcer Peter Jefferson reading from the English version of the European Union's *General Data Protection Regulation* from 2016. The Calm app also offers a wide range of comforting sounds, including "Heavy Rain", "Seashores", "Washing Machine" and "City Streets".

The most popular sounds on such sites are natural – both *biophonic* sounds from animals and plants and *geophonic* sounds like waves, rain and thunder. Other favourite categories are *technophonic* sounds from washing machines, trains and fan heaters, and white noise from televisions, as well as *anthropophonic* sounds, which are generated by people. These apps really do cater for every taste, and wisely so, for the sounds people find calming differ immensely.

The Canadian sound researcher Milena Droumeva studied a group of university students and found that, sound-wise, they preferred a coffee shop – talk, espresso machines and clattering ceramics – over a rainforest. To me, coffee shop sounds are noisy and stressful, but to the students they were familiar, predictable and therefore pleasant.

'Calm' can also be soundscapes that people design, compose, stage and share. And calm is not equivalent to an absence of sound – which is silence – so 'aural calm' also covers acoustically measurable sounds we can hear and feel.

What is more, certain input can be used to 'mask' sounds often considered unpleasant. 'White noise' is an even, constant mix of frequencies that covers our entire auditory spectrum and can hide the background hum of, say, an open-plan office. It is also used therapeutically to treat tinnitus.

'Elevator music' or 'muzak' is a type of functional

background music used by retailers and in public spaces to mask clanking, humming, beeping and talking. A typical example would be a cover version of the 1965 Beatles hit "Yesterday", arranged for panpipes and strings, and played on a synthesizer. Muzak is designed to flow into our ears as smoothly as possible, almost unnoticed. High and low frequencies are removed, along with vocals, modulations in key and any fluctuations in tempo and dynamics. Muzak is specifically designed to make shoppers and elevator passengers feel calm and comfortable, improving their experience and making customers want to linger and buy more.

I recently bought headphones that monitor my surroundings and generate a countersignal to cancel incoming noise. They can make even the loudest locations seem quiet, while excellently playing whatever sound I have selected. I use them especially when vacuuming, traveling by train or plane, or trying to concentrate while writing, in order to avoid being distracted by conversations in the next room or seagulls outside my office at Aarhus University.

Noise-cancelling headphones are another example of how technology can help us create islands of calm. Some say it would be better to reduce ambient noise instead of alleviating its consequences – and it does seem rather like putting on a gas mask to filter out air pollution instead of making sure we have clean air.

NOISE IS IN THE EAR OF THE BEHOLDER

A colleague once told me about a week-long family holiday

spent in a rented seaside cottage in North Jutland, far from the busy city. After a while they became aware of a special sound: a soft hum or distant murmur that came and went. Was it caused by traffic on the nearby highway, or by the waves of the North Sea on the beach?

Highway noise, they knew, would irritate them all week, while ocean noise would be calming, so the family collectively agreed that it must be the ocean. After that it didn't matter whether the noise came from traffic or waves. The point was that calm had descended on the cottage.

I personally find it hard to fall asleep when I can hear sounds that are beyond my control, especially music from neighbouring houses or the buzzing of an insect. If I know the neighbours will turn off the music at 2 o'clock, or if I know the bug will fly out in an hour, I have no problem. It's the uncertainty I can't stand. I actually find it calming to tell myself that the neighbours and I have agreed they will flip the switch off at 2 o'clock. That's harder to do with insects. On the other hand, I can easily take back control by getting out of bed, opening the window and shooing the insect out, which is harder to do with neighbours.

As early as the 1970s, the Canadian sound researcher and composer R. Murray Schafer argued that the ubiquity of invasive sound was making people slowly lose their understanding of what it means to concentrate. He also believed that we ought to transform silence from being a negative experience into a positive one. His ideal sounds were those of nature, for instance birdsong or gently lapping waves, and even today, nearly 50 years later, many of us harbour romantic ideas of finding or creating a place where we can live in harmony with nature, like we did 'in the good old days'.

It is naive to think that healthy sound environments can be created merely by removing measurable industrial noise. Nature too is full of bangs, crashes and howls. It is more meaningful to avoid fixed concepts of 'noise', 'calm' and 'silence' and to focus, instead, on the listener's perspective. However, this also makes the study of sound far more complex.

Raucous seagulls outside my office sometimes interrupt my work. But if I hear the exact same sound mixed with breaking waves while walking on the West Coast, it becomes part of a positive natural encounter.

The sense of hearing cannot be isolated: Our perception of what we hear always blends with other sensory input as part of a holistic experience embedded in a specific situation. Even so, conscious thought processes still give us some control over which sounds we register, and whether we allow them to bother us. I have tried to think of the seagulls outside my office as a pleasant maritime backdrop for my workday. Unfortunately, it doesn't work. Their cries may just be so abrasive that they activate my built-in contingency mechanism against loud, sudden sounds.

QUIET SHOPPING IN RØDOVRE

In 2018, a Danish shopping centre called Rødovre Centrum, west of Copenhagen, won the country's annual Stillness Prize for not playing music in the atriums and walkways, and for having shops that play only hushed music, or none at all.

The prize is awarded by the national association We Love Stillness, whose mission is to put noise problems on Denmark's political agenda, promote better soundscapes and help sound-sensitive people find homes, shops and cafés that are less noisy.

You might think that people who are sensitive to sound and noise are simply difficult and neurotic. Research has shown, however, that they have specific structures in their brains that cause them to experience sound differently than most people do, whether they want to or not.

Scientists estimate that over a third of us are sensitive to sound, irrespective of gender. A WHO study points out that in Europe, noise pollution is even more detrimental to public health than air pollution. It is also well known that constant loud noise can lead to cardiovascular disease and stress. Sound-sensitive individuals typically do not mind sounds or music in general. They just want to be in charge of it themselves, and therefore dislike background music at social events or in shops and restaurants.

SOPORIFIC TORTURE, AND SOLACE

I love audiobooks, podcasts and talk radio. I get restless and feel like I'm wasting my time if I don't have a good story or interesting talk radio in my ears as I cycle to and from work, jog or exercise, walk the dog or clean the house.

I adore my smartphone and headphones, which enable me to tap into my preferred sound content, from whatever provider, at my discretion. Such customised accessibility gave rise to the word *pod*-cast, which plays on the abbreviation of audio content that is 'personal on demand', as opposed to traditional '*broad*-cast' programmes. Podcasting lets each user personalise their consumption – location, content and timing – unlike traditional flow radio, which transmits certain content at certain times.

I usually listen to audiobooks as I wind down before bedtime. Actually, I have this idea that I can't fall asleep without one. It has become a routine. I've always identified the sound of calm with good narrator voices. I can override the mental noise that easily takes over at the end of the day by focusing my attention on a story instead. At the same time, the sound of the voice gives me a sense of comfort, safety and, along with that, calm, which can no doubt be traced all the way back to my childhood.

This shows that the defining features of an audiobook or a radio debate go far beyond their content. The narrator or radio host can employ the tone, rhythm and phrasing of the spoken word to create a mood – an ambience. Incredible as it may sound, I have even fallen asleep to an academic book about torture, the deeply disturbing contents of which are by no means sleep-inducing per se. My descent into oblivion was brought about by the voice of the narrator, John Christian Langeberg. With a deep, slightly gravelly quality and few unanticipated fluctuations, he phrases the sentences in long, smooth arches that he brings to rest at each full stop by reducing the tempo and the tone of his voice. If I am already tired and lying in a dark room, I gradually grow less conscious of the words themselves and can let a voice like his lull me to sleep.

Another favourite of mine is the opening passage of the Swedish writer Astrid Lindgren's first book about Pippi Longstocking, which introduces us to a nine-year-old girl who lives all alone in a dilapidated house on the outskirts of a tiny village. The passage may sound like the backdrop for a tragic story about a poor little orphan, but it has a very different effect on me. This is not only because it goes on to express Pippi's incurable optimism about her own situation – describing how lucky she is, being able to decide for herself when to go to bed and when to eat goodies. It is also because the Danish version of the book is narrated by Thomas Winding – whose calm, reassuring voice-overs and show hosting captivated a whole generation of Danish children. The Swedish audiobook is narrated by the author herself. With a bit of effort, speakers of the three Scandinavian languages can understand each other, and when I enjoy Astrid Lindgren's vocal storytelling, her light-footed, lively phrasing leads me on, pleasantly serving up the sentences and briefly pausing at each full stop. Both narrators, each in their own way, immediately invite the listener to curl up in front of their mental fireplace and get ready for a good story in good company. Of course Pippi is going to be all right, and so are we.

ONE FOR THE ROAD

Publishers and libraries agree that audiobooks have seen a veritable explosion in terms of numbers published, sales and library loans. For years, experts predicted that the e-book would revolutionise the publishing world, but the digital audiobook now seems to hold pride of place.

The revolution has been a peaceful one, however, with no trumpets or fanfares. This is partly because audiobooks have a long history of serving a narrow target group: those with trouble seeing or reading. Nowadays, thanks to technological developments – in particular the smartphones we often keep close at hand, day and night – audiobooks have quietly slipped into many people's lives, along with podcasts and other mediated audio content.

A digital audiobook takes up no physical space and is easy and flexible to use. Just think of the Russian writer Leo Tolstoy's epic *War and Peace*, published between 1865 and 1869 and later recorded on no fewer than 119 vinyl LPs. In the 1980s it was downsized to 45 cassette tapes, then to 50 CDs. Today, thanks to the MP3 format, the *War and Peace* audiobook is lighter than a feather, portable and far more user-friendly, although its classic contents are as weighty as ever.

This success is not attributable to technological advances alone. As a medium, the audiobook is exceptionally well suited to a busy, modern lifestyle, which often involves spending quite a bit of time commuting and exercising.

The audiobook sets our eyes, our hands, our entire body free, enabling us to 'read' a book as we run outdoors or at the gym, or travel from A to B, thereby filling the breaks in our optimised lives with something worthwhile.

This may sound like yet another stressful, efficient reorganisation of our restless lives, aimed at cramming even more things into less time – which certainly has little to do with calm.

But perhaps the ability of audiobooks to create focus and calm is one of the very reasons they have become so popular. Even as our attention latches onto the narrative, a good narrator's voice brings us evenly paced phrases and mental grounding in the here and now. Reading with our ears brings us a very different kind of calm than reading with our eyes, and in many ways it is like listening to music.

In other words, when it comes to peace of mind, listening to audiobooks in our free time is a paradox. On the one hand, it fills the gaps in our daily lives and allays our worries about time wasted. On the other hand, the audiobook is a tool to find focus and calm when we're feeling restless or impatient. Listening to literature may even give us food for thought.

As part of my research on audiobooks, in 2013 a colleague and I visited a filling station at a transport hub near the Danish–German border at Padborg, where a local library had set up a special service for a very particular target group: long-haul drivers. Using a catchy no-nonsense slogan in Danish – *Kør og hør*, or "drive and listen" – the local library loaned out CD audiobooks to long-distance drivers. Southbound, they could take one free of charge and return it, often many weeks later, on their northbound journey. This was a popular service for years, until public libraries and commercial platforms began to offer free or low-cost e-book and audiobook streaming.

Several haulage drivers told us that they normally found it hard to get through a whole book and rarely visited a library, but on the road they found audiobooks a welcome diversion that made time pass more quickly. We also learned that not only did they recommend books to each other, they also shared their audio-reading experiences with the staff at the filling station. The drivers were mainly interested in thrillers and crime novels, but many of them also asked for language courses to while away the hours at the wheel.

A SENSORY BACKRUB

A popular YouTube video called "~Simple Pleasures~ ASMR Soft Spoken Personal Attention" has been viewed tens of millions of times. It features Maria Viktorovna, who is originally from Russia but now lives in the United States, where she practises under the name Gentle Whispering. She is one of many 'ASMRtists', who make a living producing videos that evoke ASMR, or *Autonomous Sensory Meridian Response*, in listeners and viewers, by means of auditory and visual stimuli that trigger physiological reactions and give the user pleasant bodily sensations. Those who experience ASMR describe it as a tingling down their spine, or hair rising at the back of their neck. Some react to whispering, lip-wetting or crinkling wrappers. Others react to images of slow, deliberate movements or hands touching certain surfaces.

Maria begins with a deep sigh, before smiling and gazing intently into the camera and whispering a sincere welcome. Then she explains that the 40-minute video will offer the viewer relaxing movements, sounds and words. As she speaks her hands move fluidly, almost lovingly, around the camera.

The soundtrack is hyperdistinct, including the little extra sounds from her mouth as she speaks and the chafing of her clothing as she moves. Her face and hands are clean, neat and attractive, her lips and eyes discreetly accentuated by perfect makeup and her long blonde hair falling in soft curves about her face.

ASMR videos employ sounds and images, but the audio input, particularly the hushed voice, is crucial, which is why they are widely known as 'whisper videos'. They were also nominated for the Danish Stillness Prize 2018, alongside the winner Rødovre Centrum. Whisper videos often have a separate category in Calm and similar relaxation apps, but in China they have been banned altogether because their sensory effects are believed to be erotic.

Using human voices to create a sense of social intimacy and presence is well known among researchers of audio-based media. In terms of the technology, whisper videos are recorded using 'binaural' microphones, which are actually two microphones joined into one. This allows sound to be captured and reproduced to reflect the way humans naturally hear, with an ear on either side of our head. The result is an auditory 3D effect, which almost makes it seem as if the sound surrounds us. Binaural recording technology is decisive to a successful ASMR experience, because the quality lies not in the words themselves but in the way they are spoken, according to the physical, emotional and social qualities of the ASMRtist's voice.

ASMR videos are extremely slow-paced, with few cuts, if any, and panning shots that can take several hours. They dwell on details and cultivate the quality of the sound itself. This goes against the long-standing trend of favouring MP3, a format that 'crops' the sound to compress it as much as possible.

Based on psycho-acoustic analyses of the human ear, the American sound researcher Jonathan Sterne has argued that compressed MP3 files are designed to reproduce sound in "low-fi or mid-fi" quality in fairly noisy environments: outdoors, at the gym, or in an office with lots of background noise from humming computers. As he puts it, MP3 files are "meant for casual listening."

In contrast, binaural microphone technology creates a full, saturated sound that invites listeners to immerse themselves and focus on what they are hearing, a practice we could call 'deep listening' or even 'slow listening'. The better the recording and replay technology is, the higher the fidelity will be, and the greater the chances that the listener will achieve ASMR. In other words, better 'hi-fi' means a higher probability that the videos will work.

The physiological reactions behind ASMR are as old as the human body itself and can be prompted by sensory stimuli from the physical world around us. The new thing is that millions of people are consciously seeking out ASMR video and audio files through digital streaming services and relaxation apps specifically designed to help users find calm. Currently there are many millions of ASMR videos on YouTube offering to alleviate everything from insomnia and migraine to stress, anxiety and loneliness. User reactions suggest there is a certain effect, although researchers remain unsure about the mechanisms at work.

ASMR videos may have become so popular because we live in an age where contact with other people is mediated much more than before, and when we use digital media to try to compensate for fundamental needs that used to be satisfied in the physical realm. Never physically touching others can lead to a condition known as 'skin hunger'. Touch deprivation is a growing problem in Western cultures, giving rise to a whole new profession: 'cuddlers'. Such practitioners offer wellness services like holding, hugging, cuddling and tucking-in, all in the comfort of the client's own home. People who would find this awkward can satisfy their skin hunger by watching an ASMR video – part of a growing trend towards using mediatised physical touch.

ASMR videos are an example of how even physical touch and social contact are mediatised by digitised media culture. What is happening with this basic human need is the same thing we see happening with virtual visits to the doctor, for instance, or buying and selling corporate shares online: The need adapts to the options and limitations offered by a given medium.

'Social grooming' is a concept taken from the animal kingdom, often illustrated by primates picking at and smoothing each other's fur – not just to get rid of vermin, but to create and maintain social contacts. In the same way, digitally confident humans use social media as a form of mutual grooming, with 'Likes' and encouraging comments serving as a 'grooming tool' that generates positive attention and recognition of another person.

ASMRtists also cuddle and groom their listeners with intense, intimate sensory input, presenting close-up images and sounds, and 'touching' our ears with their binaural input. And, as is obvious from the comments under the "~Simple Pleasures~" video, Maria is digitally cuddled in return, having received some 17,000 positive statements and compliments, last time I checked. Her legion of grateful followers even includes a woman who watched the video and posted her thanks during childbirth, and a former combat soldier who had found respite from his otherwise fitful, nightmare-ridden sleep.

THE CHILLING SOUND OF SILENCE

Today, background music is an integrated part of virtually all films, but it was not always so. In the age of the silent movie, during the early decades of the twentieth century, live musicians in the cinema accompanied the moving pictures. One of their functions was to drown out the noise from the audience and the mechanical sounds from the film projector, but the music also created atmosphere and filled the space between the audience and the silent spectacle on the silver screen.

When 'talkies' became common during the 1930s, musical accompaniment was no longer needed, but within a few years filmmakers realised that without music the films lacked certain layers of mood and emotion. They reintroduced background music, which has played an important role in audio-visual narratives ever since.

In terms of filmmaking effects, silence does not necessarily create calm. In the American horror film *A Quiet Place*, released in 2018, the whole plot revolves around silence. The Abbott family lives on an old, isolated farmstead where they must conduct all activities in complete silence. The problem is – spoiler alert – that their post-apocalyptic world is full of blind monsters with hideously large, slimy, supersensitive ears that attack at the least auditory aberration caused by human activity. All family members must communicate wordlessly. Fortunately, they know sign language because the oldest daughter, Regan, is deaf-mute. They also have recourse to a nearby waterfall, a location where the constant noise of the gushing, searching water masks their speech. They finally discover that the monsters have a weakness: the high-frequency sound from Regan's cochlear implant processor – so a battle can be waged and potentially won, not with silence but with sound.

The film's lack of sound means that watching it is an intensely repressive and uncomfortable experience and makes moviegoers overly conscious of the sounds around them. I sat there myself, determined to avoid crackling a wrapper or clearing my throat, which increased the already excruciating suspense.

In one sense, *A Quiet Place* is almost like a silent movie with a soundtrack; its signature sound design makes the audience acutely aware of the farmstead's creaking floorboards, or a paraffin lamp tipping over.

In his 1977 book *The Tuning of the World*, R. Murray Schafer argues that we as humans like to make sounds to remind ourselves that we are not alone. We fear the absence of sound, just as we fear the absence of life and of light. The visual equivalent of silence is darkness. Without light there is darkness; without sound there is silence. Neither is absolute, of course, because in our sensory landscape the world is never pitch black or totally quiet.

A person placed in a room that is designed to be completely free of sound – an 'anechoic chamber' – will still be able to hear the whoosh of their own pulsating blood. People who have been in such an echo-free space describe the experience as terrifying, perhaps because its ultimate darkness and ultimate stillness remind us of death, a fact of life we prefer not to contemplate. Life is sound and motion, tingling vibrations and sensations. And that is why absolute silence is not even remotely calming.

NOISE CONTROL

SHHH ...

A few years ago I was at an international conference in Iceland. At the closing reception, which was held in a fairly small, low-ceilinged room, we had a stunning view of the fjord around Akureyri, our host town. Regrettably, the acoustics were ill suited to 50 people conversing, many using foreign languages with varying confidence. We all had to speak just a bit louder to be heard, which eventually turned the reception into an unintended shouting match.

A colleague of mine came over and suggested an impromptu experiment: Just for fun, why not try to 'shush' everyone until the room fell silent? That might reset the decibel level. Obviously, it would have been an extremely awkward thing to do, so we didn't. But with just a moment's calm we could have deflated a sound balloon fit to burst.

Daily life would be quieter, mind-wise and sound-wise, if people felt less compelled to communicate with each other, especially in acoustically challenging spaces. Admittedly, no communication would mean no conversations, no music, no texts, no ads and no social media, and humans don't communicate merely to convey information. We also do it to create commonality, cohesion, identity and entertainment, and to socially groom each other. Our societies simply cannot function without communication and the channels that mediate it.

Before the spread of printed books and pamphlets, people generally had to be near each other to hear spoken words or read body language, facial expressions and gestures. This put a natural limit on the scope and pace of communication. Historically, however, since the fifteenth century or so we have continuously developed new communication channels that are better and faster, forever seeking ways for our messages to reach more people in less time.

As early as the 1960s, the Canadian thinker Marshall McLuhan prophetically spoke of the world as a "global village" where virtually all information was accessible to everyone, thanks to electronic media like radio and television. The digital media culture of the twenty-first century has made McLuhan's observations more relevant than ever. Now, for better and for worse, every single inhabitant in the global village can raise their voice using images, sound and text, communicating across space and time. As in Akureyri, it's hard to get through when everyone has to shout louder and louder to be heard. We risk not being heard at all when our voices drown each other out.

Digitisation has also led to rationalisation. Today, many radio hosts must work their own audio mixers, and university researchers must often manage their own accounting. Everything from shopping to filing for divorce can be done online, and we interact physically with fewer and fewer people. But such changes in the name of efficiency have not given us more time to immerse ourselves and find calm. Instead, workplaces keep expecting us to get even faster and more efficient. More and more people find it hard to keep up and are incapacitated by stress. Maybe it's high time for a great big collective 'shush', so we can all get some peace and quiet for once.

SOLID AS A ROCK

A key scene in the Swedish comedy-drama *Force Majeure*, from 2014, shows a family of four at an outdoor table at a packed ski resort, chatting and enjoying a magnificent view of the snow-clad French Alps. Then two loud 'booms' trigger an avalanche, which ominously moves in their direction.

The father reassures the others that the avalanche experts have things under control, but as the roiling snow approaches, he bolts to save himself. The mother stays behind, holding their children. The avalanche, losing momentum, stops in the nick of time, and no one gets hurt. The rest of the film is about how the father, and the couple, try to come to terms with his gut reaction, which both he and his family find hard to justify.

The film begs the question: How would I react in such a situation? Many of us probably cannot deny that we might react just as the father did. One attempt to rationally defend his actions could be: 'He fled to save himself so he could come back and help after the mayhem.' Indeed, as the flight attendants say: "Please put on your own oxygen mask before assisting others."

Presumably, most of us hope we would be able to stay stoically calm and collected in a catastrophe, although it is hard to control the reactions evolution has encoded in our DNA to ensure our survival. The concept of 'stoic calm' stems from the philosophical school of Stoicism, which was founded in ancient Greece around 300 BC and soon spread to the Roman Empire and the rest of the world.

The Stoics practised how to stay dispassionate under pressure and in crisis, for instance by pondering death and other types of loss beyond human control. This, they believed, would prepare them to deal with such events in real life.

That does not necessarily mean we should suppress our feelings or not acknowledge them. It means that with practice we can learn to tame and work with our emotions, keeping them from counterproductively taking control. This is the direct opposite of the whisper videos I described, which aim to communicate directly with our senses and completely bypass rationality and reason. Stoic philosophy, on the other hand, is a lot like modern-day mindfulness and meditation techniques, which are inspired by Buddhism and aim to attain inner harmony and mental equilibrium.

The Roman emperor and campaign commander Marcus Aurelius (121–180) – who was memorably portrayed by Richard Harris in the blockbuster film *Gladiator* – spent most of his time ferociously defending the borders of the Roman Empire against Parthians from the east and Germanic peoples from the north. What is more, in the year 166 the plague ravaged Rome, and not for the first time. But Marcus Aurelius managed to stay level-headed and handle these huge obstacles by practising Stoic thinking. We know this because, through it all, he wrote down his thoughts, reflections and mental exercises in a kind of philosophical diary he kept, widely known in English as his *Meditations*.

These very personal writings were later reworked into 12 chapters or 'books', which included this advice to himself: "Be like the promontory against which the waves continually break, but it stands firm and tames the fury of the water around it" and "Remember too on every occasion which leads thee to vexation to apply this principle: not that this is a misfortune, but that to bear it nobly is good fortune."

To follow the Stoics, we must purge our minds of thoughts about things we can neither control nor change, which can only cause mental distress. Not until we acknowledge this will we be able to find peace of mind and act based on reason – as Epictetus, one of the fathers of Stoicism, put it nearly 2,000 years ago.

Like Buddhism, Stoicism is a sort of spiritual art of living. The ultimate goal is to become a whole human being who, in turn, is part of a greater whole. In essence, Stoicism is about developing spiritually, living in harmony with the divine in oneself and becoming one with Nature and the cosmos.

Facing the avalanche, the father in *Force Majeure* has no

time to rationally, stoically analyse the situation. He reacts instinctively. Staying calm in such a situation is not easy, but what remains is the ethical dilemma: Did he do the right thing by trying to save his own life rather than risking the whole family's survival by staying with them? In social and ethical terms he should have put his partner and children first, even in the face of death.

INTROVERTS AND EXTROVERTS

Would you prefer a birthday dinner with a few friends to a big bash with music and dancing? Do you prefer a weekend curled up in your best chair with a good book, rather than doing a family brunch and having dinner guests on Saturday, then spending Sunday organising a volunteer cake sale and barbecue? Do you shun multitasking and prefer to focus on one task at a time? Would you rather express yourself in writing than orally, and listen rather than speak? If your answer to these questions is "yes", you may be part of the group described as 'introverted' – which, in Denmark, makes up 30–50% of the population.

Research shows that creativity generally requires calm, but it also shows that our need for peace and quiet varies greatly. Introverts find calm in solitude, while extroverts find it calming to be in a group. Introverts are usually better at concentrating, while extroverts more often receive public recognition, due to the tendency in Western cultures to applaud those who thrive in the limelight.

The concepts of 'introvert' and 'extrovert' were introduced by the Austrian psychoanalyst Carl Gustav Jung in the 1920s, but it would be equally fitting to call inward-looking people 'contemplative' or 'attentive'. Concepts and labels aside, pure introverts and extroverts are extremely rare. In fact, Jung said a pure type at either end of the continuum would be insane, so we are all somewhere in between.

People react differently to sensory and social stimulation. Some need lots of stimuli, whereas others feel most alive, alert and competent in more tranquil environments. But in cultures that hail extroversion, it can be hard to fulfil the need for calm. One might even suggest there is a bias against certain personality types in the Western world.

In 2017, the Danish social science researcher Asbjørn Sonne Nørgaard personality tested a range of Denmark's serving politicians: 81 from the 179-seat national parliament, the *Folketing*, and 558 from local bodies. While there is good reason to be critical of even the most widely used personality tests, Nørgaard's study rather predictably concluded that politicians are more extroverted and domineering than the average Dane. Although Danish politics seem to be dominated by a special personality type, this does not mean there are no introverted politicians. Obviously, they too can learn to appear and act in their professional capacity as self-assured and calm under fire, but it is not likely they will ever come to love that part of their job.

Gender equality is a hot topic in Western societies, including in the political sphere. Beyond a narrow focus on gender, diversity might be well served if different personality types were represented in legislative assemblies, regional and municipal councils and corporate boardrooms. I suspect all walks of life would benefit from having leaders who, besides being good public speakers, were also good listeners. Frankly, if top executives in the world's banks and financial institutions showed less appetite for risk and more forethought, we would probably see fewer scandals in these sectors.

In cultures that cultivate individualism, extroversion also feeds the likes of Facebook and Instagram. Some people like to construct a narrative around their activities and are energised when others notice and 'like' them. Other people doubt that their lives will interest anyone else and find most social media content irrelevant.

As with sound sensitivity, introversion is also attributable to certain biological features of our nervous system, some of which are genetic. In 1989 the American developmental psychologist Jerome Kagan began a long-term or 'longitudinal' developmental study that followed 500 children from the age of four months into their teens. At intervals of a few years, he subjected them to a variety of new, age-appropriate experiences. As infants and toddlers they were introduced to surprising sounds, tastes and smells. At later ages their unknowns were mainly robots and socialising with unfamiliar children. All the while, researchers observed and documented the children's body language and typical reactions.

As infants, some reacted dramatically to the smell of

alcohol, flailing their arms and legs, making faces and crying. Others were completely calm. Now, you might think the former would later exhibit extroverted traits and the latter more introverted traits. Kagan's findings showed the exact opposite.

His study confirmed that the nervous systems of the two character types react very differently, not just to social stimuli but to *all* stimuli. Infants who reacted strongly to loud popping sounds or pungent alcoholic scents turned out also to react strongly to rollercoaster rides and their first day at school. Extroverts are often described as 'social' and introverts as 'shy'. However, people with introverted personalities often enjoy social activities just as much as extroverts do. It's just that instead of 'recharging' on socialising, introverts say it drains their energy.

FORGETTING TIME AND PLACE

Yesterday evening, after conversing with the other guests in San Cataldo's common room, I returned to my own room to work a bit more on this volume of Reflections. From my office chair behind the thick walls I looked at the little kumquat tree outside my window – and there they were, again, the swirling thoughts of all the things I could do before settling down to work. I could make a fresh cup of tea, or check the latest news online, or take a shower ... Then I thought about the Stoics, which enabled me to discipline my thoughts and turn my mind to the manuscript.

Suddenly, the hoot of an owl made me aware of my

surroundings. I had lost all sense of place and time, but my watch told me a couple of hours had passed. Better still, my Word document had grown by several pages.

When we are in the right stimulation zones we perform best and feel best, existing in the here and now, and not thinking about the future or the past. The Hungarian-American psychologist Mihaly Csikszentmihalyi refers to this state of mind as *flow*.

Flow resembles calm in that it describes a state of being where we forget time and place and achieve a sense of deep contentment. We can attain flow when an activity is not too hard but allows us to direct our full attention to whatever we are doing – writing, knitting, gardening, programming or any other pursuit that fully suits our capabilities, thus balancing stimulus, routine and challenge. If this balance tips, however, calm turns into boredom, and we once again become restless or 'uncalm'.

THE PATH TO CALM

On 25 April, a national holiday in Italy, I went with some other guests of San Cataldo and a large group of *Scalesi* townspeople to celebrate Liberation Day atop a nearby mountain. We hiked up to the holy site of Santa Maria dei Monti, some 1,000 metres above sea level, along a trail known for centuries as a leg on the pilgrimage of St. Alfonso.

After a good hour and a half on steep, uneven stone steps, we reached the white Madonna statue and the site of the festivities. Initially us visitors were not quite sure from what or whom Italy was liberated after World War II. After all, the country was long an ally of Nazi Germany. We finally established that the Italian Resistance fought against the Germans in the last years of the war. There on the summit, far from civilization, we witnessed an incongruous mix of bell-chiming, fireworks and a Catholic mass in full regalia, all set to the braying of donkeys. To an outsider like me, the scene seemed utterly bizarre.

A few hours later we took our leave to wend our way down the mountain in well-spaced single file. During the descent I realised how incredibly calm and quiet we were – absorbed in the footing of each uneven step, intent only on the intimate interaction of mind and body. With our heads free of noisy thoughts, we entered a state of flow. We were too winded to speak much, but we shared the experience, warily withdrawn yet attentive to each other, lest anyone slip on the perilous trail. The walk back to San Cataldo took almost an hour, but we all agreed that the time felt only half as long.

The Danish philosopher Søren Kierkegaard would have agreed with us on the merits of walking. In 1847, in a letter to his ailing sister-in-law, Henriette, he emphasised that "the more one sits still, the closer one comes to feeling ill", after explaining how "I have walked myself into my best thoughts, and I know of no thought so burdensome that one cannot walk away from it."

On this point, Kierkegaard is back in vogue. Renowned pilgrim trails like the Camino in Spain and Shikoku 88 in Japan are full of wanderings souls. Like the nuns of San

Cataldo, original pilgrims – whether Buddhist, Hindu, Muslim or Christian – sought to clear their minds to focus on God or the spiritual realm. As modern pilgrims treading the same paths, we probably focus more on ourselves, and on natural features and local history. Even so, the sense of cleansing and calm that pilgrims found then and find now is presumably the same.

Two weeks after arriving at San Cataldo, my academic retreat is over. It is time for the group to return to the outside world. Most of us will take the same flight back to Copenhagen, and we book two taxis to the airport to avoid the unreliability of public transport. En route to Naples we pass Mount Vesuvius, which buried the cities of Pompeii and Herculaneum in volcanic ash nearly 2,000 years ago.

Looming now above one of Europe's most densely populated regions, Vesuvius has kept its cool since 1944, when it erupted and killed 26 people. Everyone here knows this active volcano can devastate the entire area in a matter of hours, if – or rather, when – it erupts again. Its tranquil, rock-solid face is unsettling: The longer Vesuvius stays quiet, the sooner it is likely to explode again. I can't help wondering: Would I react stoically if I suddenly had to escape an erupting volcano? Would I run for dear life, or help the others get away too? Half an hour later we are engulfed in the hustle and bustle of Naples Airport, and the calm is over – for now.